NATURAL DISASTERS
MEETING THE CHALLENGE

BLIZZARD AND ICE STORM READINESS

Rachel Seigel

Crabtree Publishing Company
www.crabtreebooks.com

CRABTREE
PUBLISHING COMPANY
WWW.CRABTREEBOOKS.COM

Author: Rachel Seigel

Series research and development: Janine Deschenes, Reagan Miller

Editorial director: Kathy Middleton

Editor: Ellen Rodger

Proofreader: Roseann S. Biederman

Design and photo research: Katherine Berti

Images:

Croatian Red Cross: p. 30 (top)
Flickr
Jackie: p. 19 (top)
Massachusetts National Guard: p. 6
NC Dept of Public Safety: p. 29 (insert, top left)
Tom Booth: p. 18
Washington National Guard: p. 36
Washington State Dept of Transportation: p. 30 (bottom)
woodleywonderworks: p. 16–17 (bottom)
Getty Images: Jeff Haynes: p. 35
Scott Olson: p. 38
iStockphoto: p. 7, 25 (top), 37
tumsasedgars: p. 40 (top)
NASA
Scientific Visualization Studio: p. 13 (top)
National Climatic Data Center of United States: p. 14 (bottom)
NOAA: p. 41
CIMMS, SSEC: www.nesdis.noaa.gov—Screen Shot 2020-02-07 at 4.25.34 PM
Data NCEI: p. 17 (top)
ESRI, HERE, GARMIN, FAO, USGS, EPA, NPS: p. 22
www.weather.gov—Screen Shot 2020-02-07 at 2.49.47 PM
www.wpc.ncep.noaa.gov—Screen Shot 2020-02-10 at 4.59.39 PM
Overaasen: p. 39
Shutterstock
Arthur Villator: p. 14 (center)
Lowe Llaguno: p. 13 (center & bottom)
Raymond Deleon: p. 26 (bottom)
Steve Jolicoeur: p. 5
Tom Fawls: p. 9
Wikimedia Commons
Dahoov2: p. 42
NASA's Earth Observatory, Cooperative Institute for Meteorological Satellite Studies: p. 43
NOAA: p. 8
U.S. National Archives and Records Administration: p. 19 (bottom)
WAJWAJ: p. 4
All other images by Shutterstock

Library and Archives Canada Cataloguing in Publication

Title: Blizzard and ice storm readiness / Rachel Seigel.
Names: Seigel, Rachel, author.
Description: Series statement: Natural disasters: meeting the challenge | Includes bibliographical references and index.
Identifiers: Canadiana (print) 20190231181 | Canadiana (ebook) 20190231319 | ISBN 9780778774082 (softcover) | ISBN 9780778773993 (hardcover) | ISBN 9781427125057 (HTML)
Subjects: LCSH: Blizzards—Juvenile literature. | LCSH: Ice storms—Juvenile literature. | LCSH: Emergency management—Juvenile literature. | LCSH: Hazard mitigation—Juvenile literature.
Classification: LCC QC926.37 .S45 2020 | DDC j363.34/9257—dc23

Library of Congress Cataloging-in-Publication Data

Names: Seigel, Rachel, author.
Title: Blizzard and ice storm readiness / Rachel Seigel.
Other titles: Natural disasters: meeting the challenge.
Description: New York, New York : Crabtree Publishing Company, [2020] | Series: Natural disasters: meeting the challenge | Includes bibliographical references and index.
Identifiers: LCCN 2019052926 (print) | LCCN 2019052927 (ebook) | ISBN 9780778773993 (hardcover) | ISBN 9780778774082 (paperback) | ISBN 9781427125057 (ebook)
Subjects: LCSH: Blizzards--Juvenile literature. | Ice storms--Juvenile literature. | Winter storms--Juvenile literature. | Emergency management--Juvenile literature.
Classification: LCC QC926.37 .S384 2020 (print) | LCC QC926.37 (ebook) | DDC 551.55/5--dc23
LC record available at https://lccn.loc.gov/2019052926
LC ebook record available at https://lccn.loc.gov/2019052927

Crabtree Publishing Company

www.crabtreebooks.com 1-800-387-7650

Printed in the U.S.A./042020/CG20200224

Published in Canada
Crabtree Publishing
616 Welland Ave.
St. Catharines, Ontario
L2M 5V6

Published in the United States
Crabtree Publishing
PMB 59051
350 Fifth Avenue, 59th Floor
New York, New York 10118

Published in the United Kingdom
Crabtree Publishing
Maritime House
Basin Road North, Hove
BN41 1WR

Published in Australia
Crabtree Publishing
Unit 3–5 Currumbin Court
Capalaba
QLD 4157

Contents

What are Winter Storms?

Winter Weather

January 28, 1977, seemed like any other winter day for the people of southern Ontario, Canada and northwestern New York State. Little did they expect that later that day they would be hit with what has been called the Blizzard of the Century. For the next three days, the storm continued, completely burying cars and creating drifts from 6 to 30 feet high(1.83 to 9.14 m). **Visibility** was so poor that many people were left stranded on roads and highways, at schools, and at fire halls and police stations. Snow covered small buildings and entire cities shut down. This intense winter storm left dozens dead. It happened long before people had cell phones, and before we had the sophisticated weather forecasting we have today. Once the storm ended, people were rescued. In some areas, the clean up from the storm took up to two weeks, and the cost was an estimated $300 million dollars.

Before the blizzard started, it had snowed every day since Christmas of 1976, leading to an accumulation of **4.92 FEET** *(150 cm) of snow on top of frozen Lake Erie.*

Snowdrifts outside of a house in Tonawanda, New York after the Blizzard of '77.

More than 1,300 people are killed in car crashes on snowy or icy roads in the U.S. each year.

Many Types of Storms

The blizzard of 1977 is remembered for its sudden onset and its ferocious power. It was a lesson in the importance of winter storm preparedness because most people were caught completely off guard and there was little they could do. It taught people to be more cautious when it comes to predictions of snow emergencies, and to never underestimate the weather.

Blizzards, ice storms, nor'easters, and bomb cyclones are all different terms used to describe severe winter storms depending on where and when they occur, and the strength of the storm. They can range from a steady amount of snow for a few hours to blizzard conditions and blinding snow that last for several days. They can also occur very suddenly, knocking out heat and power, and leading to drivers being stranded on the road. They are most common in the Upper Midwest and the Great Plains areas of the U.S., China, Canada, and Russia, but they can occur all over the world, and even in the tropics on high mountaintops.

Winter storms are often called deceptive killers because of their indirect effects such as **hypothermia** and car accidents. Even when the storm is over, aftereffects such as extreme cold, snow **accumulation**, and coastal flooding can create dangerous or deadly conditions for days, weeks, or months.

State of Emergency

Some winter storms are so brutal that a government will declare a state of emergency. A state of emergency is when the government imposes rules that it would not normally be allowed to, in order to keep people safe. A state of emergency, or "snow emergency," means a city, state, provincial, or federal government orders people not to drive their cars on snowy roads because the huge amounts of snow and lack of visibility make it dangerous to do so. A winter storm-related state of emergency might also mean the government orders all **non-essential** businesses or services to close, and only emergency vehicles and snowplows be allowed on the roads. There is usually a time limit to the state of emergency. For snow emergencies, that is often the time it takes to clear roads and get things back to normal.

Members of the Massachusetts National Guard's Engineer Company help clear snow from a train station after a winter storm. Making roads and transportation safe for people to get around is a priority after a storm.

CASE STUDY
The Great Ice Storm of 1998

The greatest thickness of ice from the storm was found in Farnham, Quebec, measuring **3.07 INCHES** *(78 mm).*

Every winter storm offers valuable lessons for communities and individuals, and many lessons were learned from the Great Ice Storm of 1998. The storm earned its name because of how much destruction it caused and the duration and cost of cleanup, but it wasn't just a single storm. It was five small ice storms combined into one. It struck an area stretching from northern New York State to Central Maine in the U.S. and eastern Ontario to southern Quebec, New Brunswick, and Nova Scotia in Canada.

Nearly 40 deaths were credited to the storm, and entire cities were shut down. The storm also flooded parts of Western New York when it dumped 5 inches (127 mm) of rain in a short period of time, setting daily records for that time of year. The storm was especially devastating because of the amount of ice accumulation in some parts of Ontario and Quebec, Canada. The storm dumped more than two times the average annual amount of **precipitation**. The weight of the ice caused power lines in southern Quebec to collapse. Approximately 1,500 towers were damaged and more than three million people were left without heat or power in the middle of winter—some for up to 33 days. Crews worked for 12–16 hours a day to restore power in Quebec.

The main problem was that even though Hydro-Quebec's infrastructure met government requirements at the time, some towers and lines still weren't strong enough to stand up to such an intense storm. Since then, Hydro-Quebec has spent over $2 billion to rebuild their network with thicker transmission wires and larger foundations for each tower to make it strong enough for future storms. Another difficulty was that governments also didn't anticipate a storm this devastating and the normal storm response wasn't adequate. Since then, cities have created new emergency response procedures including early risk warnings and informing people in advance of available emergency services—helping communities and individuals be better prepared for future storms.

The Great Ice Storm of 1998 was one of the worst in North American history.

CHAPTER 1

The Science Behind Blizzards and Ice Storms

Snow is a big part of winter weather, and all snowstorms need three main ingredients to form: lift, moisture, and temperatures that are below freezing.

Moist, Cool Air

Much like other storms, winter storms start with moist air that rises into the atmosphere. Air often rises with a cold front. The air cools as it is lifted and forms clouds and precipitation. Air also rises when it moves up a mountain or other geographic feature. Moisture forms when air moves across a large lake or ocean, pulling vapor from the water. That vapor cools as it rises. With freezing temperatures in the clouds and on land, the vapor forms snow or ice. If the air warms, the snow falls as freezing rain.

What's in a Name?

There are many types of winter storms, and while they all have some characteristics in common, just like snowflakes, each one is unique. Blizzard is a term that is often used to describe snowstorms, but not all snowstorms are blizzards. A blizzard is a type of supercharged snowstorm that is usually caused when a cold front and a warm front collide.

A front, which is also called a pressure system, is a weather system that separates two types of air, known as atmospheric pressure. Atmospheric pressure is the weight of air pushing down on Earth. A lot of cool air in the atmosphere creates high pressure and hot air creates low pressure. Low pressure creates weather such as rain or storms, and when it combines with the cold air created by high pressure, the precipitation becomes snow.

A cold front is the weather word used to describe the movement of a cooler mass of air into an area of warmer air.

A front-end loader is used to clear snow from a Buffalo, New York, street after a lake-effect snow storm.

Blizzards

Some blizzards can occur without any falling snow. Instead, the snow that has already fallen is blown around by strong winds of at least 12 miles per hour (19.31 kph). When a drop in temperature follows snowfall, the snow doesn't have a chance to firm up through normal cycles of melting and refreezing and can be more easily blown around. The faster the wind blows, the farther, faster, and higher the snow travels, creating whiteout conditions. This type of blizzard is called a ground blizzard. Ground blizzards are most common in the **prairies** and countryside where there aren't a lot of buildings or objects to block the wind.

The Lake Effect

People who live near the Great Lakes in North America experience a kind of storm known as a lake-effect storm. Lake-effect snow happens when cold air moves across the unfrozen, warmer waters of a lake. The cold air warms and becomes more humid as it streams across the lakes. It then cools as it rises, forming precipitation clouds that drop snow over land near the lakes. Most lake-effect storms on the Great Lakes occur between November and February, but on October 12–13, 2006, Buffalo, New York, experienced an unexpected early lake-effect storm which was nicknamed the "October Surprise." This early storm dropped 3.5 feet (142 cm) of snow in some areas of Buffalo but some surrounding areas received almost no snow at all. The snow had mostly melted by October 15.

Bomb Cyclones

The term bomb cyclone sounds frightening and brings to mind an exploding hurricane of snow. It occurs when a **low pressure system** moves in and air pressure falls quite low in a short span of time—such as 24 hours. Known to **meteorologists** as "explosive cyclogenesis" or bombogenesis, this weather system brings strong winds and snow that seems to explode when the pressure suddenly drops.

It is most often a winter ocean event. Warm air coming off the ocean and cold air above land collide, causing a major drop in atmospheric pressure. The air starts to move, and the rotating Earth creates a cyclone effect. Because of this, these storms are also sometimes called winter hurricanes.

Siberian Purga

The buran is a strong and cold wind that creates extreme blizzards in north-central Asia in regions such as Mongolia and Siberia, and even parts of Alaska. Buran winds are known to be bitterly cold with blowing ice and snow. Over the **tundra**, these blizzard winds are called purga. In Alaska, the northeasterly winter winds are known as burga. Sometimes, the buran blows as far as Italy, bringing freezing storms to areas near the Adriatic Coast.

Outside of North America, blizzards occur **ANYWHERE IN THE WORLD** *that gets snowfall such as China and Russia.*

A man walks along the street during a blizzard in a city in Siberia, Russia

Ice Storms

Most winter storms involve snow, but precipitation that falls as freezing rain can cause ice storms, which can be extremely dangerous and deadly. In addition to making roads and other outdoor surfaces as slippery as ice rinks, heavy amounts of ice can severely damage trees and knock out power lines. Just a half inch of ice (1.27 cm) can add as much as 500 pounds (226 kg) of weight to a power line and can increase the weight of a tree branch by 30 times. Strong winds add extra force to the weakened power lines and trees and increase the possibility of major damage.

Ice as a Nuisance?

An ice storm occurs when there is an ice **accumulation** of more than a quarter inch (0.64 cm) and different amounts of ice. Accumulations of less than 0.9 inches (0.25 cm) of ice are called a nuisance ice storm. When there is one quarter to one half an inch (0.25 to 1.27 cm) of ice accumulation, it is considered to be a disruptive ice storm, while widespread accumulations of over half an inch (1.27 cm) are called crippling ice storms. With this kind of storm, the damage from the storm is severe, and power outages can last from several days to several weeks.

Ice from winter storms can increase the weight of tree branches by **30 TIMES** *or more. Branches can break with just one half to one inch (1.27 to 2.5 cm) of ice.*

Ice sheets from a storm can be so heavy, they break trees and branches and even split them in half.

How, Where, and When

Ice storms begin with freezing rain, but despite its name, freezing rain isn't literally frozen rain. It starts out as snow high up in the atmosphere, and melts when it passes through a warm layer of air. As the liquid passes through the cold layer below, it begins to refreeze. If the precipitation freezes while still in the air, it becomes sleet when it lands on the ground. If it falls without freezing, the cold layer cools the rain to below freezing in a process called supercooling. When the supercooled drops hit an object or land on the ground, a layer of ice builds up.

Rimes and Glazes

Two different types of ice form during an ice storm. Rime ice is white or milky and coarse like sugar. It is often seen on trees or mountains in winter. Glaze ice is clear and smooth like an ice cube. It is harder and sticks firmly to whatever surface it forms on. Because of this, it causes more damage than rime ice, which is less dense and doesn't stick as firmly.

Glaze ice is considered the more damaging kind of ice because it heavily coats surfaces, breaking tree limbs.

Hard rime usually forms in windy conditions.

Soft rime is very fragile and falls off easily.

West Coast Storms

Winter storms are all about location. On the West Coast of the U.S. and Canada, winter storms are usually caused by an occurrence known as an atmospheric river. An atmospheric river is much like a river in the sky that gets pushed along by strong winds. An extreme atmospheric river can cause **torrential** rain, flooding, and dangerous mudslides. It also brings snow to the higher parts of the mountains near the coast.

Pineapple Express

One of the most common types of atmospheric rivers is the Pineapple Express. This weather phenomenon gets its nickname because it originates in the Hawaiian Islands, pulls moisture from the Pacific Ocean, and travels to the Pacific coast of North America. It causes heavy rain and snowfall in the mountains of the North American coast. Once it drops its moisture, the air mass blows over the mountains of the Pacific Northwest as a Chinook wind. This is a dry warming wind that causes warm temperatures and melting snow.

A series of atmospheric rivers were responsible for the strong winter storms that hit the U.S. West Coast in December 2010.

Location, Location

Winter storms can vary in location, but they are most likely to occur in any area that is likely to be affected by cold temperatures and freezing rain. In the U.S., blizzards can occur anywhere except for the Gulf Coast and California Coast, but are most common in the upper midwest and the Great Plains. The flat ground in the plains allows the wind to reach blizzard-speed, and cold temperatures make the snow light and easily blown around. In Canada, blizzards are most frequent in the prairies, eastern Arctic, and eastern Ontario.

Nor'easters

In the northeastern part of the U.S. and Canada, blizzards usually come from hurricane force storms moving over the Atlantic Ocean. They originate as nor'easters, or extratropical cyclones that develop when warm **Gulf Stream** ocean currents from the tropical regions of the Atlantic Ocean meet cold air masses swooping in from Canada. The result is a strong low pressure system. The name nor'easter comes from the direction of the winds, that blow in from the northeast. They bring hurricane-force winds along with either rain, freezing rain, or snow. They are most powerful from October through April.

How Long Do They Last?

Ice storms occur most frequently in the northeastern U.S., Ontario, Quebec, and the Atlantic provinces in Canada. Winter storms can also occur farther south, such as the 1994 ice storm that covered Mississippi in large amounts of ice and caused large-scale damage in 11 other states including Tennessee, Texas, and Georgia.

Saskatchewan, Canada

Maine, U.S.A.

A satellite image of the North American Ice Storm of 1998.

Warming Arctic temperatures are thawing permafrost, or frozen soil. As it thaws, it releases greenhouse gases from the ancient plants and animals buried in the soil. Greenhouse gases contribute to more global warming.

CASE STUDY
Global Warming and Winter Weather Changes

Blizzards in the northeast United States and eastern Canada are likely to increase as **global warming** makes Arctic temperatures rise. That's the conclusion of a team of climate researchers from Rutgers University in New Jersey. The team studied winter weather patterns back to 1950 and determined that heavy snowfalls were more frequent since 1990, and "extreme snowfalls have occurred primarily during the recent decades."

The researchers published their study in a research journal called *Nature Communications* in 2018. They found the frequency of severe winter storms in the northeastern U.S. increased with persistent warming of the Arctic since 1990. When Arctic temperatures were high, places such as Boston and New York were four times more likely to experience severe winter weather storms, and "weather bombs." The western United States saw warmer and drier weather. Researchers believe a warming Arctic leads to a more wildly moving **jet stream**.

Researchers measured air temperatures and pressures in the Arctic and compared them with readings from a winter weather severity index. The index uses data, or information, from 100 weather stations. The study analyzed 10. In that sample, it found the Arctic is warming at a faster rate than other areas of the world. Other scientists believe the phenomenon written about in the *Nature Communications* journal is also happening in other areas of the world. Some parts of Eurasia have much colder winters and are also receiving record snowfalls as Arctic temperatures warm.

CHAPTER 2

Studying Winter Storms

Impact on Humans

Snowzilla is the nickname for a January 23, 2016 nor'easter that hit the northeastern U.S., affecting 33 million people, canceling 13,000 flights, and killing 55 people in storm-related accidents. It was an historic and deadly blizzard that dropped up to 3 feet (91 cm) of snow over parts of the northeastern U.S., including New York, New Jersey, Pennsylvania, Virginia, and Washington, D.C.

Of the estimated 55 deaths, several were related to heart attacks caused by shovelling snow. Others resulted from vehicle accidents on slippery roads, and hypothermia—where people were out in the weather and froze to death. The extreme cold temperatures, heavy snow, ice, and high winds of severe winter storms means they can be deadly and their dangers are often underestimated.

Cold on Skin

Extended exposure to cold can also cause frostbite and hypothermia. Frostbite is a freezing of the tissue underneath skin and occurs most often in the fingers, toes, nose, ears, cheeks, and chin. Hypothermia occurs when a body loses heat faster than it can produce it, causing body temperature to drop to dangerous levels. When your body temperature drops, your heart, organs, and nervous system don't work properly and can lead to death. Limiting your time in the cold and dressing in warm layers are key to staying safe and preventing injury or death.

"SNOWZILLA" *ranks among the* TOP 10 WORST *blizzards in U.S. winter weather history.*

Total storm snowfall (inches)

4 10 20 30 50+

The Storm of the Century dumped up to 6 inches (15 cm) of snow in the Florida panhandle and 6 foot (1.8 m) snowdrifts near Birmingham, Alabama—both areas where snow is rare.

Cost of Winter Storms

The damage caused by winter storms can be very costly. A heavy buildup of snow and ice can bring down trees, power lines, and communication towers. Communications and power can be disrupted, and travel by air and by ground is often delayed or canceled. The cost of cleanup after the storm, as well as the cost of lost sales from closed businesses, can be high. The January 2016 blizzard (Snowzilla) was estimated to have cost the economy $350–850 million in lost business and wages. Since 1980, the National Oceanic and Atmospheric Administration (NOAA) has recorded 16 winter storms with damage of more than a billion dollars.

Storm of the Century

People like to give severe storms nicknames that reflect their size or impact. The Storm of the Century that hit the Eastern Seaboard in March 1993 is the most expensive storm to date, costing $9.2 billion dollars in damage across the U.S. The storm closed almost all highways from Atlanta and to the northwest of the city. Every major airport on the East Coast was shut down for the first time in history, and from Florida to Maine, 10 million people and businesses lost power due to high winds and fallen trees.

Societal Impact

Winter storms can completely paralyze a region or a city. That doesn't just inconvenience people, it makes it difficult for people to access health care, such as hospitals—even in emergencies. A January 18, 2020 blizzard shut down the island of Newfoundland in Canada, and the government ordered a state of emergency in the capital city of St. John's. Wind gusts of 103 miles per hour (167 kph) and 30 inches (76 cm) of snow created whiteouts in many areas. People had to dig out from snow-blocked front doors. Some worried that they would be unable to pick up needed prescriptions for health conditions. The armed forces were called in to help people dig out after the blizzard.

Without Power and Heat

Power outages, loss of telephone service, and shutdowns or delays in transportation systems are common during severe winter storms. As the safest place to be is home, these storms can also isolate people in their homes, cutting them off from friends, family, and normal activities. This can have a negative effect on mental health—particulary among the elderly and people with disabilities.

Holiday Shut Down

A severe winter storm that hit Northeast China between January 25 and February 6, 2008, occurred right before the Chinese Spring Festival (Chinese New Year) holiday. The storm stranded hundreds of thousands of travelers, many of whom were migrant factory workers who were returning home for the holiday to visit their families. Thousands of passengers were stuck on delayed electric trains and at airports after canceled flights. Highways were closed because of the snow, and bus travel was canceled due to icy roads. Heavy snow caused a delay in coal supply which accounted for more than 80 percent of China's electricity generation. Many people were trapped in their homes and left without power, heat, or light. Overall, an estimated 60 million people across the country were affected by the storm that hit six provinces.

As many as 500,000 people were left stranded at the train station in the southern city of Guangzhou after a power failure in the neighboring Hunan province disrupted train service.

Communities Affected

The aftermath of a winter storm can have a lasting effect on a community, especially when it isn't prepared. On February 5, 1978, a nor'easter caused hurricane force winds, record snowfall, and coastal flooding in Massachusetts and other parts of New England. Boston received a record 27.1 inches (68.83 cm) of snow. It can take years for communities to recover financially from an unusually severe storm. Preparation requires warning of a storm and the resources to withstand it and clean up.

No Warning

One of the problems in the blizzard of 1978 was a lack of advance knowledge of the storm's severity. Local weather forecasts at the time had a reputation for being inaccurate, so people didn't believe them. By the time they realized the storm was dangerous, many people were stranded at work and school, and some couldn't get home for several days. Since then, storm forecasting has become more precise. Communities have also invested in better plowing equipment and have emergency plans that set out when to close roads, which services to make a priority, and when to call a state of emergency.

More than 3,500 cars were found abandoned and buried in the middle of roads during the cleanup after the storm.

Snowmelt

After a heavy snowfall, if the weather warms up rapidly, the snow and ice on the ground can melt too quickly for it to be absorbed by the frozen ground. When the meltwater has nowhere to go, it can lead to flooding. Flooding is also an **ecosystem** hazard because large quantities of water can negatively affect plant and animal habitats, and their food supply.

Melting Disaster

In March 2019, a record-breaking bomb cyclone brought hurricane force winds and heavy snow and blizzard conditions to the central United States.

A sudden warm-up and deep snow that melted quickly combined with heavy rains and frozen ground caused flooding in the U.S. Great Plains. As a result, many roads and bridges in Iowa, Nebraska, and Missouri were damaged.

The disaster was caused by a combination of weather, water, a changing climate, and the failure of dams and **levees** which normally prevent lakes and rivers from flooding. All of these factors demonstrated the need to better plan for the changing climate and extreme weather to prevent similar future disasters.

Many communities, homes, and businesses in Iowa and other parts of the midwestern U.S. were literally underwater as a result of the bomb cyclone. A levee break can flood an entire town.

Knowledge is Power

In the U.S. weather watches, warnings, and advisories are issued by the National Weather Service. This is a federal government agency that provides forecasts and warnings to the public. As part of the National Oceanic and Atmospheric Administration (NOAA), it provides near and long-term weather forecasts, records of past weather for researchers to study, and education and safety advice. The NOAA's Weather Ready Nation program is targeted at helping Americans prepare for changing weather conditions.

Working Together

Because the U.S. has the largest variety of extreme weather events, the NOAA felt it needed to help the public be better aware of risks. Weather Ready Nation brings weather experts and researchers together with emergency planners. The goal is to improve community resilience—or the ability to recover quickly from difficulties. They do this by using the most timely and accurate forecasts and better emergency response to weather events so they don't become disasters. Weather Ready Nation also works with community organizations that help people recover from weather-related disasters. For example, Save the Children, a charity that helps children, provides disaster preparedness training and activities for children in kindergarten to Grade 5.

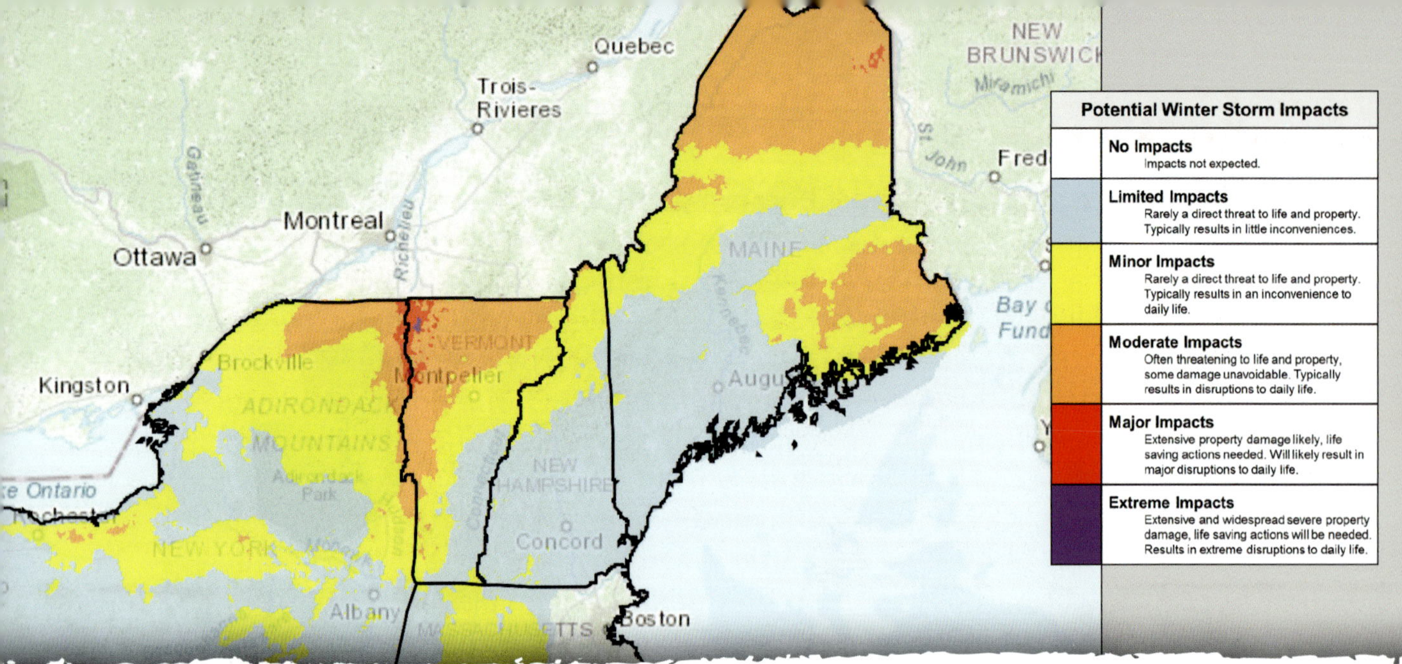

Meteorological Service

In Canada, the Meteorological Service of Canada, and Environment Canada, are government organizations that provide weather forecasting and warnings. The Meterological Service has storm prediction centers throughout the country. Environment Canada tracks weather, collects data, and provides weather alerts.

A Winter Storm Severity Index (WSSI) is a tool that weather forecasters use to determine how severe a storm is and how to tell the public about the storm's potential impact. It's not a forecast and not meant to be the only source of information on a storm.

The impacts of a winter storm can be felt long after the storm is over. Scientists created the Northeast Snowfall Impact Scale (NESIS) to help rate the severity of a winter storm in the northeastern U.S. immediately after it occurs. The scale was developed for the Northeast because of the economic and transportation impacts a storm can have. The impact scale has five levels of intensity, similar to the scale used to measure hurricanes and tornadoes, and ranks storms from one (notable) to five (extreme). Where the NESIS differs from other meteorological tables is that it considers factors such as geography and population as well as weather factors such as snow and wind speed. The Regional Snowfall Index (RSI) is an evolution of the NESIS which includes the impact of snow on other regions of the U.S. It's used for major snowstorms that affect the eastern two-thirds of the U.S.

The blizzard of 1996 was a major nor'easter that deposited 4 feet (1.2 m) of snow on the U.S. east coast between January 6 and January 8. Philadelphia received 30.7 inches (78 cm) of snow, most of which fell in 24 hours. It was the most of any city in the storm's path. The mayor declared a state of emergency and only police and emergency workers were allowed to drive on the streets. The storm was one of the most extreme to ever hit the region and is just one of two ranked extreme on the NESIS and one of 26 since 1900 on the RSI.

SCIENCE BIO

National Severe Storms Laboratory

A long time ago before the invention of computers, the only way to predict the weather was through your own experience. People might have looked to weather **folklore** or to nature to predict what was to come. This was not very accurate, and many people were caught out in storms or blizzards because they didn't have warning. Today, meteorologists use data that includes ocean temperatures, land and ice conditions, and atmospheric information to make their forecasts.

Many scientists study weather and climate in laboratories, including the National Severe Storms Laboratory (NSSL). The NSSL is a federal organization under the National Oceanic and Atmospheric Administration (NOAA), which is working to improve the amount of lead time and the accuracy with which severe weather is forecasted. Their research covers extreme weather such as tornadoes, flash floods, lightning, and winter weather.

To better predict whether or not a winter storm will develop, the NSSL uses dual-polarization radar. Radar is a detection system that uses radio waves to detect the position of things in the distance and the direction of moving objects. Dual-polarization radar sends out two different radars which are separated horizontally and vertically. These provide a two-dimensional picture of the size and shape of the object that is falling, and help meteorgologists identify rain, hail, snow, and ice pellets.

The NSSL also uses a set of step-by-step computer instructions called Q2 that detect mixed precipitation, which is a combination of rain and partially melted snow known as sleet. These tools allow forecasters to more accurately predict winter storms because it gives them more information about what to expect.

Sleet can be as big a threat as snow because it can freeze into solid ice within a few hours of falling and is extremely difficult to remove.

CHAPTER 3

Meeting the Challenge

Accurately predicting winter storms is a difficult process, because conditions in the atmosphere and on the ground can vary in different areas. Even a small temperature change can affect whether precipitation falls as rain or snow. Those differences create different types and amounts of snow in various cities or neighborhoods.

Blizzard prediction relies on temperature predictions and computer weather models. A weather model combines data about current weather conditions such as wind speed, wind direction, air temperature, pressure, and humidity with information about how weather behaved in the past to form mathematical equations. The calculations formed by the computer help us understand what kind of precipitation will occur.

Understanding the local climate is also important when it comes to accurately predicting winter storms. Weather only describes the conditions in the atmosphere from day to day while climate describes weather conditions over a period of 30 years or more. Knowing what kind of conditions are usual for the region makes it easier to predict what kind of weather will occur over the long term and short term.

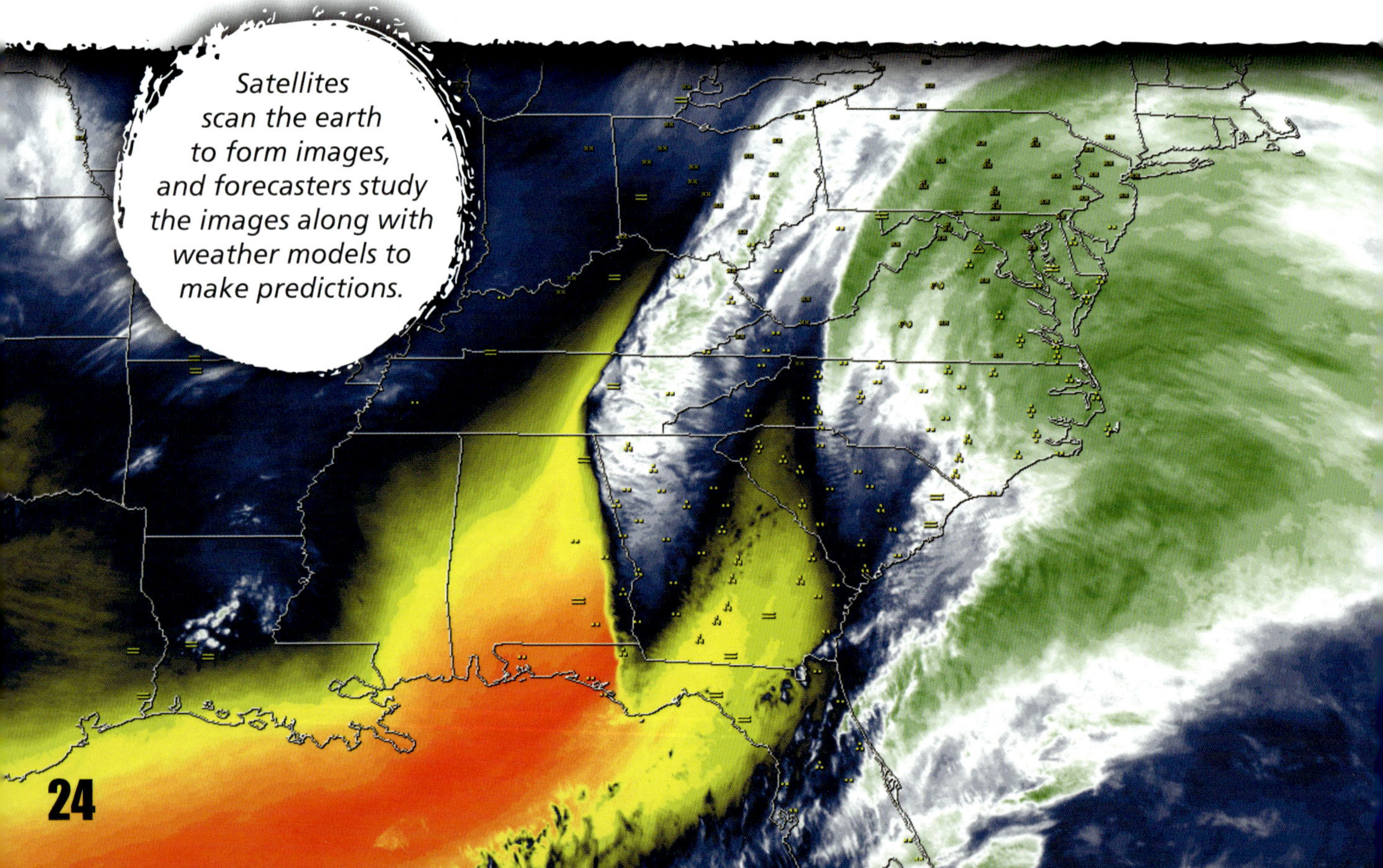

Satellites scan the earth to form images, and forecasters study the images along with weather models to make predictions.

The National Weather Service's ASOS system detects weather changes and automatically sends information on changes out through its weather networks. It also provides weather information to airports and aircrafts.

A weather station is a device that collects data related to weather and the environment using multiple sensors.

Tracking and Monitoring

Both the U.S. and Canada have special government agencies that are in charge of providing weather forecasts and warnings to citizens. In the U.S. this is the National Weather Service (NWS) and in Canada it's the Meteorological Service of Canada.

The NWS monitors storms with the use of Automated Surface Observing Systems (ASOS). The ASOS is a system of radar sensors that reports the weather every hour as well as special weather observations if conditions suddenly change. The ASOS is used in flight planning and in-flight safety because weather conditions can affect aircraft safety while flying.

Meteorologists also use a wide network of observational systems, such as satellites and Doppler radar, to watch the storm and figure out its direction. Satellite images are useful tools for figuring out the cloud patterns that cause storms and how they are moving. Doppler radar is a special kind of radar that can determine the strength of the precipitation as well as wind speed and direction. This is helpful when forecasting near mountains or lakes or oceans because it identifies the presence of ingredients needed for precipitation. If the radar shows wind blowing up the mountain, forecasters know there will be lift. If it shows wind blowing over a large body of water such as a lake, it tells them there will be moisture.

Tracking Storms at Home

Thanks to modern technology, there are many ways that individuals can track winter storms from their phones or on their computers.

Websites such as CNN's winter storm tracker, the National Weather Service radar map, AccuWeather's live radar map, and the Weather Channel's storm tracker app allow people to track winter storms from their computers and their phones. On the weather.gov website, there are full forecasts, up-to-the-minute weather alerts, and county-by-county weather reports.

Cell phone apps such as Dark Sky and iWeather offer live radar, and notifications about alerts, conditions, and forecasts. Local newspapers, broadcast TV stations, and Twitter feeds provide official updates from local weather organizations.

Local weather stations can provide storm information specific to your area.

According to the Insurance Bureau of Canada, nearly all catastrophic losses of over **$25 MILLION** in the last 30 years were weather related.

SCIENCE BIO

Weatherlogics, Winnipeg, Manitoba

Matthieu Desorcy and Scott Kehler are obsessed with weather. Both men grew up on the Canadian prairies where extreme weather was a normal part of life. Summer storms could bring hail and winter could last eight months a year. Both men studied atmospheric science in school, and Kehler spent summers chasing storms and tornadoes in the Canadian prairies and the northern U.S. Both understood that weather impacts everything from agriculture to transportation. This led them to set up Weatherlogics, a company that offers clients sophisticated weather information including hail storm tracking and pavement temperatures to help them make better informed decisions.

One of their services is an exclusive road weather system called Road Weatherlogics that maps highway conditions and identifies specific areas where there could be hazards such as blowing snow and freezing rain, or where roads could be covered with slush, snow, or ice. This information is especially helpful to trucking companies, which can use the information to make safe decisions about routes.

While most weather websites can only provide current road conditions, Road Weatherlogics provides an hourly forecast of road conditions for the next 48 hours. By having advance information, drivers can plan based on what's ahead rather than on the current information.

The company also provides information about road temperatures, which helps cities make key decisions about snow plowing and ice prevention and removal.

Storm Warnings

When a winter storm is coming, letting people know what to expect is essential to allow them time to prepare. In order to warn people that they are at risk the NWS uses specific terms. The term used depends on how much snow falls in a 12 to 24 hour period.

The amount of snow required to issue a warning depends on snow amounts in the area. A winter storm watch is issued at least 24 hours in advance of a storm. It means that there is at least a 50 percent chance that it will occur, but the exact location and timing aren't known yet.

Winter Storm Advisories

When the winter storm is less than 24 hours away, the watch becomes either a winter storm warning or a winter storm advisory. A warning means that the storm conditions can be life threatening, can cause damage to your property, and that travel will be difficult if not impossible. Winter storm warnings are issued when heavy snow of at least six inches (15.24 cm) in 12 hours is expected or 8 inches (20.32 cm) in 24 hours. A warning can also be when freezing rain is expected to produce a large and possibly damaging amount of ice or when blizzard conditions are expected.

A winter storm advisory is issued when between three to five inches (7.62 to12.7 cm) of snow or when ¼ inch (0.635 cm) freezing rain/sleet is expected within 12 hours.

Advisories help governments prepare road clearing fleets for potential storms. Preparation saves lives.

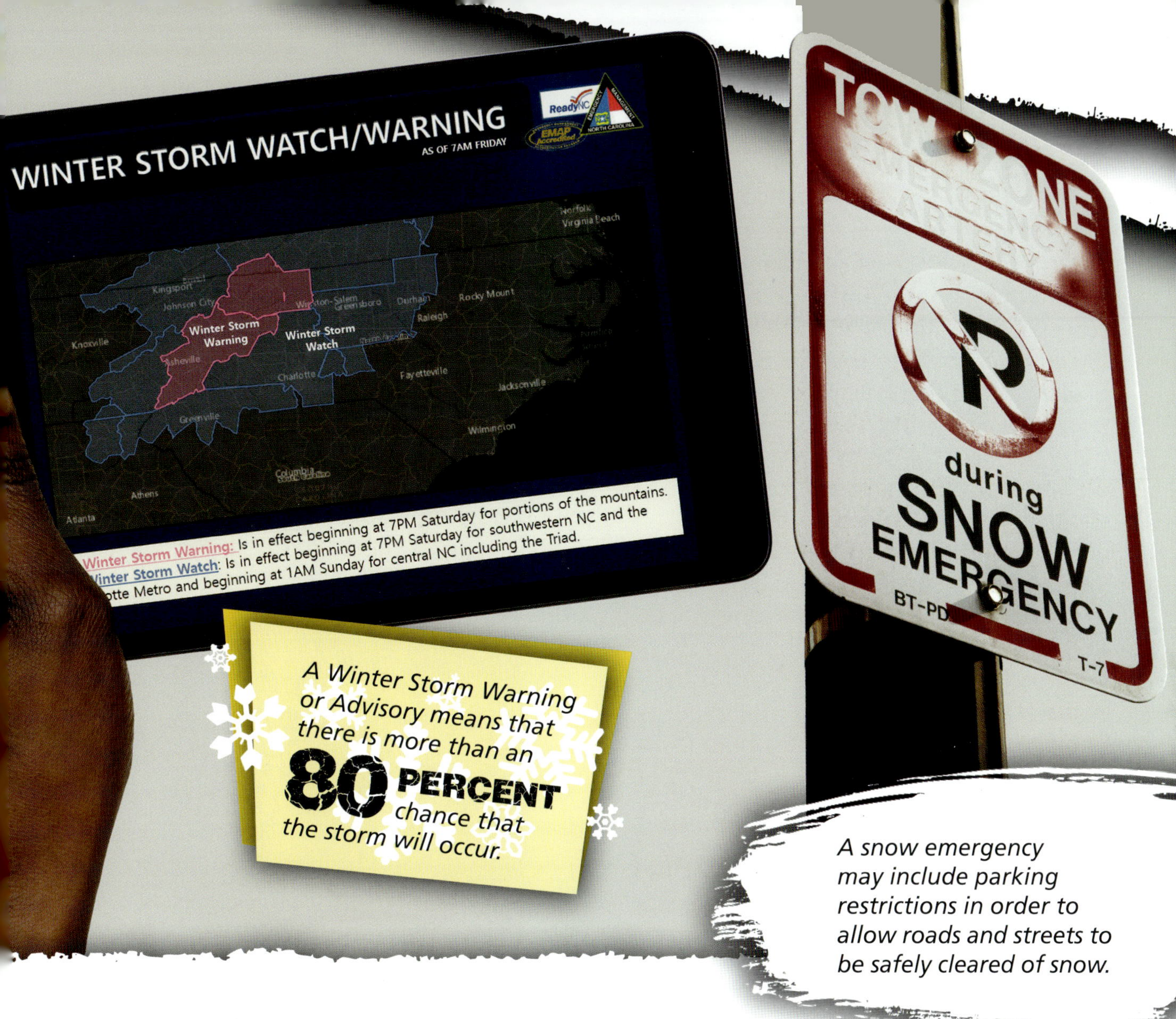

A Winter Storm Warning or Advisory means that there is more than an **80 PERCENT** chance that the storm will occur.

A snow emergency may include parking restrictions in order to allow roads and streets to be safely cleared of snow.

Disaster Response

A snow emergency happens when a snow storm is so severe it is dangerous to keep roads, schools, and businesses open. When local governments declare snow emergencies, non-essential services are shut down. These may include airports, schools, and public buildings. Hospitals, polices stations, and fire departments need to be kept open. The goal is to ensure the safety of people by keeping them home and off the roads.

Snow emergencies are often leveled 1 to 3, with each level representing an increasing hazard. Level 1 means blowing and drifting snow on roads with a caution to drive carefully. Level 2 means only drive when necessary, and level 3 means roads are closed to all but emergency vehicles. Each local government has its own rules that it follows for snow emergencies. The rules also include what streets are plowed first. Parking rules may also go into effect to allow plows to clear the streets and to let emergency vehicles through.

State of Emergency

If the storm becomes life-threatening, a city, state, or province may declare a state of emergency. In October 2019, the Canadian province of Manitoba declared a state of emergency after an early winter storm dropped more than 1.64 inches (50 cm) of snow in some parts of the province. The storm was extremely powerful and arrived before the leaves had fallen off the trees. As a result, the weight from the combination of rain, snow, and wind caused an abnormal amount of damage to trees, transmission lines, poles, and power lines. The state of emergency gave police, firefighters, and forestry workers the ability to work on private property where many trees and power lines had fallen. It also allowed Manitoba to request help from other provinces and states to help repair the damage.

In addition to the actions taken by cities, states, or provinces, there are many agencies such as the Red Cross and in the U.S., the Federal Emergency Management Agency (FEMA) that assist with disaster response.

The Red Cross is an organization that helps people all over the world recover from disasters. Here, workers in Croatia are bringing supplies to people in isolated areas after a heavy snowfall.

In February 2019, the Governor of Washington declared a state of emergency after a series of winter snowstorms pounded the Pacific Northwest.

The Baltimore Humane Society asked the community for help removing ten downed trees on the property.

Snow removal crews in Baltimore worked in **12 HOUR SHIFTS** and used over **300** pieces of snow-removal equipment to clear primary and secondary roads after the storm.

Disaster Recovery

When the storm is over, recovery can be a slow process. The timeline for recovery depends on the community and how severe the storm was. One of the most important aspects of recovery is making sure that communities have plans in place to recover from the storm, and especially for cleanup when the snow stops falling.

The immediate step that cities must take to put their communities on the road to recovery is to dig out the city and get transportation systems moving again. The January 2016 blizzard dumped 29 inches (74 cm) of snow in Baltimore and Washington and was expected to take days to clean up because of the volume of snow. The city hired additional private contractors to assist with snow removal and disposal.

Debris removal and cleanup is another important part of the process. The stronger the storm, the more likely there is to be damage. A fast cleanup gives people an immediate sign of improvement and goes a long way to building resilience.

Recovery Challenges

Even after the snow stops falling, there are many problems caused by the snow and ice that make recovery a challenge.

One of these is the risk of sewer back-up and flooding after a rapid snowmelt. Large amounts of water caused by snowmelt can overload the drainage system and cause the sewer to back-up and send the dirty water back into the building along with harmful bacteria. Ice dams also cause a risk of flooding. An ice dam is a ridge of ice that forms at the edge of a roof and prevents snow from draining off the roof. This causes the water to back up and leak into ceilings and walls and rot the roof boards.

The other big challenge to recovery is repairing the damage to infrastructure caused by heavy snow, ice accumulation, and high winds. They destroy power lines, damage telecommunications systems, disrupt water supply, shut down major roads, and collapse buildings. No matter how well prepared a community is, there is still likely to be damage, and it can take days or weeks to repair.

Cities such as Seattle that don't get regular snowfall during the winter also face challenges because they don't buy a lot of equipment that is dedicated to snow removal. This means that when they do get snow, removal takes longer than in a city such as Buffalo that receives regular snow. Cities also face additional challenges in removing snow from side streets and rural areas. It can often take several hours or even days to remove snow from all of the streets in a city. Since clearing major streets where traffic is heavier usually takes priority, side streets and rural areas can remain unplowed for several days.

Cars parked in the street are an obstacle to cleanup crews after a winter storm.

Infrastructure damage
Ice dam
Sewer drain with snowmelt water

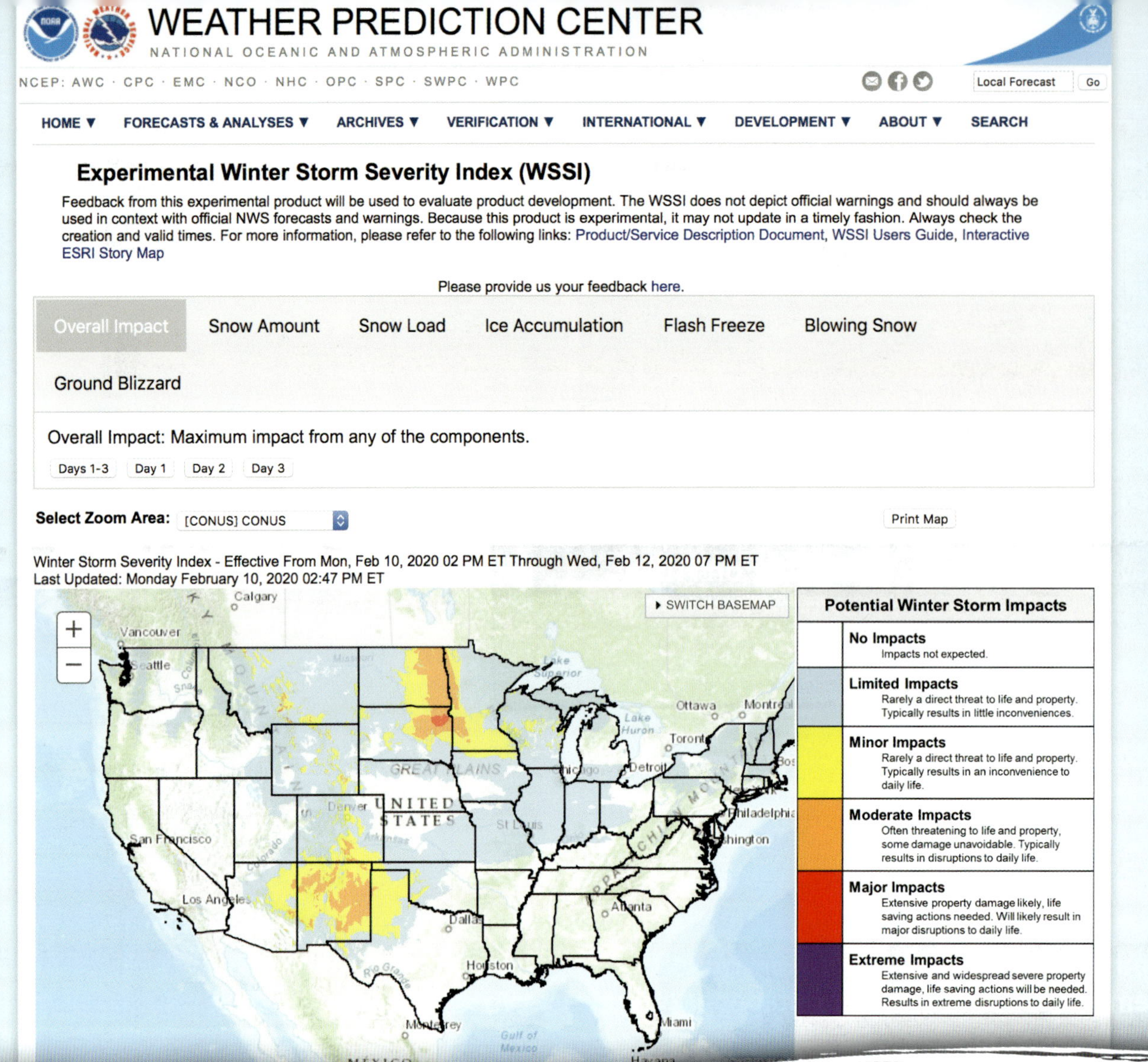

By using colors, the WSSI allows users to understand threats from winter storms just by looking at it.

New Ways of Studying Winter Storms

Scientists are always looking for new ways to track and monitor winter storms with better accuracy. The National Weather Service introduced an experimental new tool called the Winter Storm Severity Index (WSSI) to track snow and ice severity.

Most weather watches, warnings, and advisories only look at total accumulation but don't consider that severity is more than just amounts. The goal of the WSSI is to provide clues about what level of severity the winter precipitation will reach, and how it will impact society. It combines multiple types of data about how the snow and ice accumulation impact transportation, power lines, trees, and individual areas. Those conditions are then color coded according to the severity. No color means there is no expected impact from the storm while dark red indicates an extreme impact.

CASE STUDY

Super Bowl Ice Storm 2000

The icy roads caused hundreds of car wrecks, including a **47 CAR PILEUP** on Interstate 20.

In January 2000, all of North Georgia was hit by two severe ice storms in a week that paralyzed parts of Atlanta. The first storm occurred on Sunday, January 23, a week before the city was set to host the Super Bowl. Less than one week later on Friday, January 28, a second storm hit just two days before the Super Bowl.

Atlanta is not a city that experiences many snowstorms. Winter is usually pretty mild, so when the storm hit, they were incapable of handling it. To make matters worse, none of Atlanta's news outlets had correctly predicted the weather, and the state was caught off guard. Georgia's emergency systems had trouble coping with the increased demand caused by power outages and traffic accidents. Atlanta, its northern suburbs, and the entire North Georgia area was covered in ice ¼ to ½ inch (6 to 13 mm) thick. Falling branches from trees brought down power lines, and at least 675,000 homes were without power. The governor declared a state of emergency for some districts.

Having learned from the mistakes of 2000, when another major storm hit in 2014, they were better prepared. The state brought in 180 tons (163 metric tons) of additional salt to melt the ice, and was considering redirecting traffic to help clear the most heavily used roads and highways. Hundreds of National Guard troops were ready to evacuate hospitals and nursing homes, and the president declared a state of emergency.

The Super Bowl continued but many nearby vendors were affected as events and activities were either reduced or restricted.

CHAPTER 4

Facing Future Winter Storms

Changing Winter Storms

When scientists talk about climate change, people usually think it means warmer weather, but that is not the case. Global warming, which is the warming of the earth has made weather more extreme and unpredictable. As temperatures rise, the number of extreme weather events will increase in frequency, and the strength and impacts will get worse. It also means that wet areas will get wetter, and dry areas will get drier.

Snowfall depends on moisture in the atmosphere, and climate change is causing warmer temperatures which leads to more water evaporation. More water evaporation means that there is more water available for precipitation. In some areas, the warmer temperatures increase the chances of that precipitation falling as rain. In colder areas where temperatures stay below freezing, this could actually cause an increase in snow. In the last 50 years, the frequency of snowstorms in the eastern two-thirds of the U.S. has increased.

In the last 20 years, the number of blizzards each year has **DOUBLED** *from an average of* **NINE** *from 1960–94 to* **19** *since 1995.*

Climate change is increasing and intensifying snowstorms. Many areas require the help of the armed forces to clean up after these storms.

Planning for Disaster

Winter is guaranteed to bring winter storms and they can become a disaster quickly and unexpectedly. Effective disaster **mitigation** measures or hazard prevention is the foundation for disaster management. The major components of mitigation are preparation, response, and recovery. These are the actions taken by a community to prevent the risk of threat before it happens and then to respond to and recover from its effects.

Risk avoidance from winter storms includes protecting buildings and structures from snow, ice, and wind. In New York City, regulations are placed on how buildings are designed and maintained. Construction regulations require all new buildings to have roofs that can hold the weight of snow. Windows also have to be **insulated** against extreme cold to prevent heat from escaping or cold getting in. Informing and educating the public about the impacts of winter storms is also part of risk avoidance and can help reduce the number of injuries or deaths related to the storm.

Winter storm response is a big part of lessening storm impact. One of the greatest challenges faced by communities is coordinated planning and response. This allows different groups to make informed decisions about the potential impact of winter storms on people, the economy, infrastructure, and the environment.

The weight of snow and ice can cause roofs to collapse and can kill or injure the people inside.

Gaps in Response and Readiness

No matter how well a community plans for a winter storm, there are still gaps in response.

For people with disabilities, winter weather can be a struggle, forcing them to risk injury or stay inside. Many cities don't have the design or the structure in place to make it possible for people with disabilities to get around in winter. Clearing streets for cars becomes a priority, while sidewalks, footpaths, and stairs are left uncleared. Cities need to focus on land use and city design to make cities more accessible.

Blizzards and ice storms can also affect mass transit systems such as trains, subways, and buses, and block roadways and highways, which makes travel difficult or impossible. While underground portions of subways are not usually affected by weather, snow and ice can cause signal switches to freeze and block tracks, delaying or stopping service.

Winter weather can cause overcrowding and delays for trains in cities when tracks need clearing or equipment is affected by the cold.

Heavy snow in Norway makes snow removal a challenge. Norwegians use a superpowered snowblower to remove snow. The snowblower is designed for clearing large areas such as airports. It has a 2,000 horsepower engine and can remove 10,000 pounds (4,536 kg) per hour.

Removing the Snow

Heavy snowfall and winter storms can create challenges in snow removal and storage. In most cities, snow and ice removal is made more complicated by parked cars or cars abandoned on highways. Plows are forced to try and clear around the cars. Many cities have bylaws that prevent on-street parking during snowstorms.

Finding a place to put the snow is also a challenge after a heavy snowfall. A common method of removing snow is to haul it away in dump trucks, deposit it in snow dump sites, and leave it to melt naturally. Luckily, countries with high snowfall are working on technological solutions to make the process of snow removal and storage easier and more cost effective.

In Japan, the Hokkaido Regional Development Bureau invented an automated robotic snowplow named Yuki-Taro in 2005. Yuki means snow in Japanese and Taro is a popular boy's name. The device has a pair of video cameras built into the eyes and an internal Global Positioning System (GPS) that uses satellite information to help it detect obstacles in its path. The front shovel scoops up the fallen snow and reduces it into 28.5-pound (12.93 kg) bricks.

Japanese cities are also using stored-up solar power from the summer to heat roads in the winter by circulating hot water below the pavement to melt snow.

Green Innovations

Several million tons of road salt are used each winter to help melt snow and ice, but as the snow and ice melt, the salt gets washed into rivers and lakes, polluting water. It's also damaging to roads and concrete structures because it speeds up **corrosion**. As a result, some cities such as Chicago and Calgary (a city in Alberta, Canada) are using sugar beet juice to treat roads before snowfalls. Roads treated with beet-extract are less likely to get a buildup of ice, and it is more environmentally friendly than salt.

The company Solar Roadways, based out of Idaho, is developing solar technologies that can tolerate the weight of a car and be used on roads, driveways, patios, or anywhere that snow and ice accumulate. This could help keep roads free of snow and ice during winter and create safer surfaces.

Canada's Earth Innovations Inc. developed EcoTraction as a pet-friendly and green alternative to salt. The mineral used is a **porous**, soft, honeycomb-structured stone that acts like a sponge absorbing slippery water on the surface. Unlike salt, EcotTraction doesn't leak into the ground. Instead, it keeps its spiny structure on the ice and snow creating instant traction for **pedestrians** and vehicles, helping to prevent slips and falls. Over time, the sharp edges crush and break down the ice, allowing it to be swept away.

In the United States, road salt is spread at a rate of
10,000,000
tons (9,071,847 metric tons) per year.

When the sugar molecules from beet juice are mixed with salt, it melts the ice at a lower temperature than salt alone, and is less damaging to roads and the environment.

From Sunday through Thursday after the storm, fire crews made **1,300 RUNS** which is **57 PERCENT** above average for that time of year.

CASE STUDY

Winter Storm Goliath

Winter Storm Goliath was a disastrous two-part winter storm that hit the southern plains of the U.S. from December 24 through December 29, 2015, and was the deadliest storm system of the year in the U.S.

The first round of precipitation affected parts of the West, and a section of the Upper Midwest, from December 24 through December 26. The second round fell over the southern plains bringing a blizzard that lasted from late December 26 through December 27. From December 27 through December 29, snow and ice from the second storm spread into the central and northern plains, the Great Lakes region, and parts of the Northwest.

One year after Goliath, Lubbock, Texas, made some major changes to how they handled winter storms. The storm brought more than 12 inches (30.48 cm) of snow in just 48 hours, and major roadways became impassable for days. The following year they purchased two new trucks, with attachable snowplows and created an emergency snow removal plan for priority streets. The streets were determined based on how close they were to hospitals, fire trucks, and major parking lots. The city also made several important changes to their emergency preparedness plans, which included a focus on communication and the creation of a weather resource guide. After Goliath, they realized they didn't do enough to keep the citizens informed, and the guide will put all of the key information about the city's response and services in one spot.

CHAPTER 5

Conclusion

Most people are likely to experience a winter storm during their lifetime. Whether it be strong winds or ice or heavy snow, they can be extremely damaging, knocking out power and blocking roads and highways. There is no question that winter storms are changing. Climate change has been linked to an increase in the intensity and amount of snowfall in winter storms. Climate scientists have observed an increase in heavy snowfalls, and in cities, extreme snowfalls have primarily occurred in the last couple of decades. The impacts of the storm can affect a community for days, weeks, or months, and have demonstrated the importance of preparedness and mitigation.

Today we are much better prepared for winter storms than we were even ten years ago, and scientists continue to study winter storms and find ways to make them less destructive. Forecasters are now able to predict and monitor winter weather earlier and more accurately, and information is available in real time through computers, cell phones, and tablets. New technologies and green solutions are giving communities new ways to handle snow removal that are more cost effective and better for the environment, and structural improvements are being made to buildings to make them better withstand snow and ice.

It is impossible to entirely prevent winter storms from happening, but with proper planning and preparation, and an understanding of the risks, we can minimize their impact. Studying past disasters offers valuable insight into how communities can improve their disaster response and plan for the future.

The worst blizzard in U.S. history was the Great Blizzard of 1978, which struck the Ohio Valley and Great Lakes regions.

The President's Day storm of 2003 was a record-breaking East Coast blizzard that lasted from February 14–19.

Ask Yourself This

Based on what you've learned in this book, what are some of the ways humans have learned from past winter storms and their damaging impacts over the years?

1. How and why do you think winter storms are changing?
2. Do you think that humans are better prepared for winter storms than they were ten years ago? Explain why or why not.
3. How does technology help cities respond to and recover from winter storms? Provide examples.
4. Why is accurately predicting winter storms challenging?

Bibliography

Introduction

"2018–19 U.S. Winter Storm Season." TornadoGenius Wikia. https://bit.ly/385Ng2A

"Ice Storm." Awstats.net. www.awstats.net/ice-storm

"Lessons Learned from Ice Storm of '98" CBC News. January 09, 1999. Accessed August 11, 2019. https://bit.ly/2S244Sx

"Natural Hazards: Snow & Hail Storms." Natural Hazards—Snow & Hail Storms. www.n-d-a.org/snow-hail-storm.php

"People Who Died Directly or Indirectly from a Blizzard or Snowstorm." Genealogy Project. https://bit.ly/2SoLWRI

"Winter Storm." Science Daily. www.sciencedaily.com/terms/winter_storm.htm

Chapter 1

Bartels, Meghan. "What Is the Pineapple Express? Extreme Rain May Cause Flooding in California." *Newsweek*, March 21, 2018. https://bit.ly/39fwxKB

"Ice Storms: Hazardous Beauty." The Weather Doctor. Accessed August 11, 2019. https://bit.ly/2Sq72iG

Sawe, Benjamin Elisha. "Where Are Blizzards Most Likely To Occur?" WorldAtlas, August 29, 2018. https://bit.ly/2txzuHb

The Weather Channel. "5 Things To Know About Ice Storms." The Weather Channel. November 14, 2018. https://bit.ly/2UsiSvt

"What Is a Blizzard?" LiveScience. https://bit.ly/31vq0sE

"What Is a Bomb Cyclone?" Mental Floss, March 13, 2019. https://bit.ly/2UDSCyv

"What Makes A Snowstorm A 'Blizzard'?" *Farmers' Almanac*, January 17, 2019. https://bit.ly/2GZNDzO

"Winter Weather Types." NOAA National Severe Storms Laboratory. https://bit.ly/2RYH5I2

Chapter 2

"Avoiding the storm." Innovation.ca. Accessed November 4, 2019. www.innovation.ca/story/avoiding-storm

Baltimore Sun. "Winter Storm Cleanup: How Long Will It Take?" *Baltimore Sun*, July 1, 2019. https://bit.ly/2OwjPPt

Henry, Alan. "How to Obsessively Track This Weekend's Snowstorm." Lifehacker, January 21, 2016. https://bit.ly/2UrKGjF

"Government's Role." Blizzard Preparedness. Accessed August 18, 2019. https://bit.ly/3beysRu

"Manitoba to Declare State of Emergency Due to October Snowstorm" CBC News. October 13, 2019. https://bit.ly/37XboEF

"Ohio Committee for Severe Weather Awareness Snow Emergency Classifications." Ohio Committee For Severe Weather Awareness. Accessed November 3, 2019. https://bit.ly/382jOe1

Robbins, Liz. "After Blizzard Ends, a Slow Recovery." *The New York Times*, December 27, 2010. www.nytimes.com/2010/12/28/nyregion/28blizzard.html

Shepherd, Marshall. "Four Lessons From 'Bomb Cyclone' Flooding In The Great Plains." *Forbes Magazine*, March 18, 2019. https://bit.ly/2uqglHB

"Winter Watches and Warnings." WeatherWorks, February 3, 2016. https://bit.ly/2v40IFQ

"Winter Weather." NOAA National Severe Storms Laboratory. www.nssl.noaa.gov/research/winter

"Winter Weather Detection." NOAA National Severe Storms Laboratory. https://bit.ly/2Oq9chh

"Winter Weather Response." Seattle Department of Transportation. https://bit.ly/394UmEG

Chapter 3

"Avoiding the storm" Innovation.ca. Accessed November 4, 2019. www.innovation.ca/story/avoiding-storm

"Government's Role." Blizzard Preparedness. https://bit.ly/2S27CEl

Henry, Alan. "How to Obsessively Track This Weekend's Snowstorm." Lifehacker, January 21, 2016. https://bit.ly/3bokOeC

"How to Clean Up Safely After a Snowstorm" MetLife, February 28, 2018. https://bit.ly/2SmywWE

"Manitoba to declare state of emergency due to October snowstorm" CBC News. October 13, 2019. https://bit.ly/2UtegFx

"Ohio Committee for Severe Weather Awareness Snow Emergency Classifications." Ohio Committee For Severe Weather Awareness. Accessed November 3, 2019. https://bit.ly/382jOe1

"Winter Watches and Warnings." WeatherWorks Your Weather Experts. February 3, 2016. https://bit.ly/31uIk59

"Winter Weather." NOAA National Severe Storms Laboratory. www.nssl.noaa.gov/research/winter

Chapter 4

Anzilotti, Eillie. "How To Make Snow-Bound Cities Less Of A Frozen Hell For People With Disabilities." Fast Company, January 11, 2018. https://bit.ly/371i1Vd

"Climate Change and Extreme Snow in the U.S." National Climatic Data Center. https://bit.ly/39fkKf1

Dotray, Matt. "One Year after Winterstorm Goliath, Lubbock Says Lessons Learned." *Lubbock Avalanche*, December 26, 2016. https://bit.ly/39bNJAq

McNally, Victoria. "Remember That Time Japan Invented The Most Adorable Robot Snowplow Ever?" The Mary Sue, January 27, 2015. www.themarysue.com/japan-snow-wall-e

Miller, Brandon. "As the Arctic Warms, Nor'easters May Get Worse." CNN, March 21, 2018. https://cnn.it/2GU3AaO

Nosowitz, Dan. "11 Ways Science and Technology Are Waging War on Winter." *Popular Science*, January 26, 2015. https://bit.ly/2SjcBj7

Simon, Matt. "The Controversial Link Between Epic Storms and a Warming Arctic." *Wired*, March 13, 2018. https://bit.ly/2OsQPIE

"Winter Weather—How to Manage the Risk?" 2019 NYC Hazard Mitigation. https://bit.ly/2S3vAzb

Learning More

Books

Black, Vanessa. *Ice Storms*. Jump!, 2019.

Bow, James. *Snow and Ice Storms*. True North, 2019.

Honders, Christine. *Chasing Extreme Weather*. PowerKids Press, 2019.

Keppeler, Jill. *Blasted by Blizzards*. PowerKids Press, 2018.

Websites

Ready is a national public service campaign designed to educate and empower people to prepare for, respond to, and minimize impact of disasters.
www.ready.gov/kids/know-the-facts/winter-storms-extreme-cold

Website with weather information for kids.
www.weatherwizkids.com/weather-winter-storms.htm

Website with weather information for kids.
www.weatherwizkids.com/weather-safety-winter-storm.htm

National Center for Atmospheric Research weather site for kids.
eo.ucar.edu/kids/dangerwx/blizzard3.htm

Winter Weather Basics from National Severe Storms Laboratory.
www.nssl.noaa.gov/education/svrwx101/winter

Glossary

accumulation The increase or growth of something

corrosion The breakdown of materials due to chemical reaction

ecosystem All of the plants, animals, and other living things that make up the communities of life in an area

folklore The traditional beliefs, customs, and stories of a community passed through generations by word of mouth

global warming The increase in Earth's average temperatures that is causing corresponding climate change

Gulf Stream A warm ocean current flowing north from the Gulf of Mexico that affects weather

hypothermia A medical condition that happens when your body loses heat faster than it can produce it, causing body temperature to drop dangerously below 95 degrees Fahrenheit (35 °C)

insulated To be protected from heat, cold, or noise

jet stream Narrow or meandering air currents in the atmosphere that bring winds that affect weather

levee A bank built along a river to prevent flooding

low pressure system A weather system where air pressure is lower than surrounding areas and is associated with hurricanes and other weather

meteorologist A type of scientist who studies the atmosphere to predict and understand Earth's weather

mitigation Acting to reduce the severity or difficulty of something

non-essential Something not absolutely necessary

pedestrian A person who walks along a road or a developed area

porous An item that is full of tiny holes or openings so liquid can pass through

prairies Ecosystems made up of temperate grasslands and savannas

precipitation The release of water from the sky in the form of a liquid or solid such as rain, sleet, or snow

torrential Pouring quickly and in large quantities

tundra The vast treeless plains of the Arctic

visibility The quality or degree to which a person can see

Index

About the Author

Rachel Seigel is an avid book enthusiast with over 15 years of experience working with schools and libraries matching books to readers. She is also the author of several non-fiction books for children. When she isn't writing or researching fun facts, she enjoys spending time with her boyfriend and her mischievous dog.